Level 2

Mathematics
3rd Edition

Nancy McGraw

Simple Solutions Learning, Inc.
Beachwood, OH

Summer Solutions

Level 2

Mathematics

3rd Edition

All rights reserved. No part of this publication may be reproduced or transmitted in any form or by any means, or stored in a database or retrieval system, without prior written permission from the publisher, unless otherwise noted. Reproduction of these materials for an entire school or district is prohibited.

Printed in the United States of America

ISBN: 978-1-934210-34-5

United States coin images from the United States Mint

Cover Design: Randy Reetz
Editor: Randy Reetz

Copyright © 2018 by Bright Ideas Press, LLC
Beachwood, Ohio

Instructions for Parents/Guardians

- *Summer Solutions* is an extension of the *Simple Solutions* Approach being used by thousands of children in schools across the United States.

- The 30 lessons included in each workbook are meant to review and reinforce the skills learned in the grade level just completed.

- The program is designed to be used three days per week for ten weeks to ensure retention.

- Completing the book all at one time defeats the purpose of sustained practice over the summer break.

- This book contains lesson answers in the back.

- This book also contains a "Who Knows?" drill and Help Pages that list vocabulary, solved examples, formulas, and measurement conversions.

- Lessons should be checked immediately for optimal feedback. Items that were difficult or done incorrectly should be resolved to ensure mastery.

- Adjust the use of the book to fit your summer schedule. More lessons may have to be completed during some weeks.

Summer Solutions Level 2
Mathematics

Reviewed Skills Include

- Addition and Subtraction With Regrouping
- Place Value
- Fact Families
- Rounding to Hundreds
- Basic Multiplication Facts 0, 1, 2, 5, 10
- Basic Geometry
- Simple Fractions
- Charts and Graphs
- Math Vocabulary
- Telling Time and Finding Elapsed Time
- Counting Money
- Simple Measurement Conversions
- Word Problems

Help Pages begin on page 63.

Answers to Lessons begin on page 69.

Lesson #1

1. 456 + 285 = ?

2. Draw 2 congruent circles.

3. What will be the time 20 minutes after 6:00?

4. 540 − 239 = ?

5. Write the symbol that tells you to multiply.

6. How many quarts are in a gallon?

7. 8 ÷ 2 = ? (Hint: How many groups of 2?)

8. A ladybug weighs more than a pound or less than a pound?

9. 434 ◯ 443

10. What fraction of the figure is shaded?

11. 9 × 3 = ?

12. How much money is shown?

13. Which digit is in the ones place in 742?

14. Ellen bought a pen for $1.50 and a notebook for $1.65. How much money did Ellen spend on these supplies?

15. How long is the pencil?

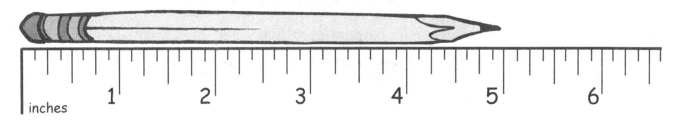

Summer Solutions© Mathematics — Level 2

1. $\begin{array}{r}\overset{1}{4}56\\+285\\\hline 741\end{array}$	2. OO	3. 6:20
4. $\begin{array}{r}5\overset{3}{\cancel{4}}\overset{}{0}\\-239\\\hline 301\end{array}$ $\begin{array}{r}5\overset{3}{\cancel{4}}0\\239\\\hline 301\end{array}$	5.	6.
7.	8.	9.
10.	11.	12.
13.	14.	15.

Lesson #2

1. What time is shown on the clock?

2. 77 + 35 = ?

3. A butterfly has 4 wings. How many wings are on 5 butterflies?

4. Find the mode of 27, 50, 13, 41, and 27.

5. Is 648 an even number or an odd number?

6. What is the name of the shape?

7. Write 341 using words.

8. There are _____ inches in a foot.

9. 0 × 2 = ?

10. How much money is shown?

11. 924 − 551 = ?

12. The product is the answer to a(n) _____ problem.

13. 14 ÷ 2 = ? (Hint: how many groups of 2?)

14. In the number 807, which digit is in the tens place?

15. What temperature is shown on the thermometer?

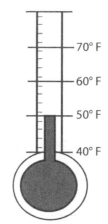

Summer Solutions© Mathematics   Level 2

1.	2.	3.
4.	5.	6.
7.	8.	9.
10.	11.	12.
13.	14.	15.

Lesson #3

1. $1 \times 5 = ?$

2. Is the line on the figure a line of symmetry?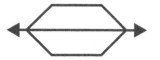

3. $554 + 386 = ?$

4. Write the symbol that tells you to subtract.

5. Ricardo planted 3 rows of bean plants. Each row had 5 plants in it. How many bean plants did Ricardo plant?

6. Count by twos. 56, _____, _____, 62, _____

7. Round 41 to the nearest ten.

8. Draw 2 squares that are not congruent.

9. Does a bag of groceries weigh more than or less than a pound?

10. Write the name of each shape.

    A)   B)   C)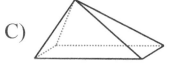

11. How much time has passed from one clock to the next?

12. $3 \times 5 = ?$

13. $94 - 29 = ?$

14. Which even numbers come between 31 and 35?

15. How long, in inches, is the pen?

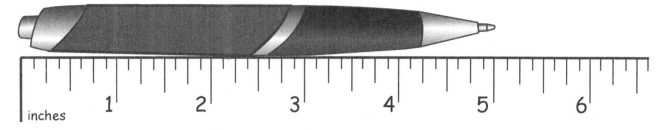

Summer Solutions© Mathematics — Level 2

1.	2.	3.
4.	5.	6.
7.	8.	9.
10.	11.	12.
13.	14.	15.

Summer Solutions© Mathematics   Level 2

Lesson #4

1. Which is greater, 5 nickels or 3 dimes?

2. 316 ◯ 163

3. Write the first 5 odd numbers.

4. 718 + 256 = ?

5. How much money do you have if you have 8 nickels?

6. The answer to an addition problem is called the _____.

7. Which digit is in the thousands place in 7,419?

8. Does a dictionary weigh more than a pound or less than a pound?

9. A party hat has the shape of which solid?

10. Write the time shown on the clock.

11. 659 − 164 = ?

12. Round 37 to the nearest ten.

13. Choose the number that comes between 135 and 230.

   A) 415   B) 121   C) 205

14. 0 × 2 = ?

15. How many fish are in the lake?

   How many more catfish than bass are in the lake?

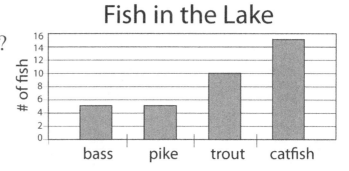

8

Summer Solutions© Mathematics — Level 2

1.	2.	3.
4.	5.	6.
7.	8.	9.
10.	11.	12.
13.	14.	15.

Lesson #5

1. Which is longer, 3 weeks or 1 month?

2. 34 + 12 + 25 = ?

3. What is the name of the shape?

4. 2,356 ◯ 1,735

5. Two figures having the same size and the same shape are _____.

6. 2 × 2 = ?

7. Find the mode of this set of numbers. 10, 14, 18, 10, 21

8. Write the time shown on the clock.

9. 753 − 439 = ?

10. Which digit is in the hundreds place in 5,207?

11. What fraction of the rectangle is shaded?

12. How many quarts are in a gallon?

13. How much money is shown?

14. 3 × 5 = ?

15. Mrs. Thomas' house has three floors. There are 6 rooms on the first floor, 4 rooms on the second floor, and 5 rooms on the third floor. How many rooms are in Mrs. Thomas' house?

Summer Solutions© Mathematics — Level 2

1.	2.	3.
4.	5.	6.
7.	8.	9.
10.	11.	12.
13.	14.	15.

Lesson #6

1. 89¢ − 57¢ = ?

2. Write a fact family for 4, 5, and 9.

3. Draw two congruent hearts.

4. Is the number 389 even or odd?

5. 664 + 267 = ?

6. 5 × 5 = ?

7. Does this show a line of symmetry?

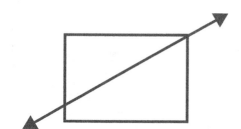

8. Which digit is in the thousands place in 8,631?

9. What fraction of the rectangle is not shaded?

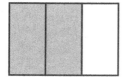

10. 2,527 ◯ 2,257

11. Is a couch about 6 inches long or 6 feet long?

12. Fill in the missing numbers. 260, 270, ____, 290, ____

13. Round 66 to the nearest ten.

14. How many inches are in a foot?

15. July 1st will be on which day of the week?

   The play is 2 weeks from June 6th. What is the date of the play?

June

S	M	T	W	T	F	S
		1	2	3	4	5
6	7	8	9	10	11	12
13	14	15	16	17	18	19
20	21	22	23	24	25	26
27	28	29	30			

Summer Solutions© Mathematics   Level 2

1.	2.	3.
4.	5.	6.
7.	8.	9.
10.	11.	12.
13.	14.	15.

Lesson #7

1. Write the even numbers between 41 and 45.

2. 653 − 337 = ?

3. How much money is shown?

4. How many minutes are in an hour?

5. 6 × 2 = ?

6. Write the time on the clock.

7. 53 + 98 = ?

8. How many days are in 2 years?

9. What is the name of this shape?

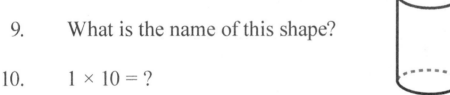

10. 1 × 10 = ?

11. The answer to a subtraction problem is the _____.

12. Write the month that comes before December.

13. Which number comes just before 1,000?

14. Write 797 using words.

15. There were 482 people at the water park on Thursday. On Friday, 875 people visited the park. How many more people were at the water park on Friday than on Thursday?

Summer Solutions© Mathematics   Level 2

1.	2.	3.
4.	5.	6.
7.	8.	9.
10.	11.	12.
13.	14.	15.

Lesson #8

1. $8 \times 2 = ?$

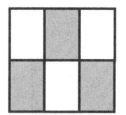

2. What fraction of the square is shaded?

3. There were 4 boats on the lake on Saturday. Each boat had 5 people aboard. How many people were in all four boats? Use a drawing to help you.

4. How many months are in 2 years?

5. Write the name of the shape.

6. How much time has passed from one clock to the next?

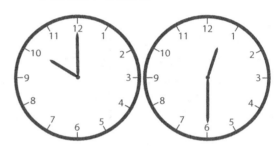

7. $642 + 353 = ?$

8. List the odd numbers between 70 and 76.

9. $6 \div 2 = ?$

10. The product is the answer to a(n) _____ problem.

11. $530 - 319 = ?$

12. Which shaded part is greater?

13. How much money is 8 quarters?

$\frac{1}{6}$ ◯ $\frac{1}{4}$

14. Round 35 to the nearest ten.

15. Which class is the longest? How long is it?

    What time does math class start?

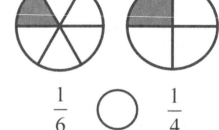

Class Schedule	
8:30–9:00	Spelling
9:00–10:00	Reading
10:00–10:30	Math
10:30–11:00	English

Summer Solutions© Mathematics — Level 2

1.	2.	3.
4.	5.	6.
7.	8.	9.
10.	11.	12.
13.	14.	15.

Lesson #9

1. What is the mode in this set of numbers? 24, 17, 31, 10, 31

2. Write the time shown on the clock.

3. 10 × 5 = ?

4. 988 ◯ 1,000

5. What number follows 1,346?

6. A foot has how many inches in it?

7. Which digit is in the tens place in 4,075?

8. 3 × 2 = ?

9. How much money is ten dimes?

10. Round 485 to the nearest hundred.

11. Is the number 7,427 even or odd?

12. There are _____ quarts in a gallon.

13. A cellphone is about 5 inches long or 5 feet long?

14. Draw a rectangle and shade $\frac{2}{5}$ of it.

15. Write the temperature shown on the thermometer.

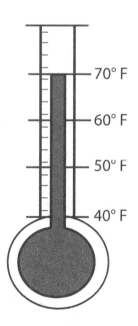

Summer Solutions© Mathematics — Level 2

1.	2.	3.
4.	5.	6.
7.	8.	9.
10.	11.	12.
13.	14.	15.

Lesson #10

1. Which is longer, 8 months or 1 year?

2. 514 + 293 = ?

3. Draw 2 congruent ovals.

4. How much money is shown?

5. 0 × 2 = ?

6. Round 264 to the nearest hundred.

7. 828 − 464 = ?

8. What time is it?

9. 5,324 ◯ 4,254

10. Does this show a line of symmetry?

11. Which digit is in the hundreds place in 8,064?

12. 346 + 428 = ?

13. What fraction of the pie is shaded?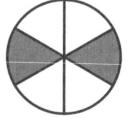

14. Put the numbers in order from least to greatest.

    465   1,254   787   325

15. How many children chose dogs?

    How many more children chose hamsters than cats?

Favorite Pets	
Pets	# of Children
dogs	|||| ||||
cats	||||
hamsters	|||| ||
fish	|||

Summer Solutions© Mathematics   Level 2

1.	2.	3.
4.	5.	6.
7.	8.	9.
10.	11.	12.
13.	14.	15.

Lesson #11

1. The number that appears most often in a set is the _____.

2. 7 × 5 = ?

3. Round 396 to the nearest hundred.

4. 69¢ + 53¢ = ?

5. Write the name of the shape.

6. Count by twos.   62, 64, _____, 68, _____, _____

7. Write the even numbers between 81 and 87.

8. What fraction of the pie is not shaded?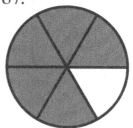

9. 957 − 279 = ?

10. Which digit is in the ones place in 5,328?

11. Are these shapes congruent?

12. Count by fives.   45, _____, 55, _____, _____

13. What is the time shown on the clock?

14. Write a fact family for 4, 9, and 13.

15. Leon ate 9 cookies on Monday. He ate 11 cookies on Tuesday and 13 more cookies on Wednesday. How many cookies did Leon eat during the three days?

Summer Solutions© Mathematics   Level 2

1.	2.	3.
4.	5.	6.
7.	8.	9.
10.	11.	12.
13.	14.	15.

Lesson #12

1. 4 × 10 = ?

2. What fraction of the rectangle is shaded?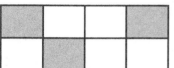

3. How much money is 6 quarters?

4. 3 + 8 = ?

5. How much time has passed from one clock to the other?

6. 518 + 233 = ?

7. There are _____ minutes in a half-hour.

8. The sum is the answer to a(n) _____ problem.

9. Write the name of this shape.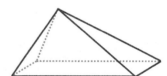

10. 835 − 452 = ?

11. 1,177 ◯ 946

12. Round 346 to the nearest hundred.

13. 1 × 5 = ?

14. Write the first four odd numbers.

15. How long is the arrow, in inches?

Summer Solutions© Mathematics Level 2

1.	2.	3.
4.	5.	6.
7.	8.	9.
10.	11.	12.
13.	14.	15.

Lesson #13

1. $518 + 364 = ?$

2. Draw a rectangle and shade $\frac{3}{4}$ of it.

3. 943 ◯ 1,500

4. Find the mode of 27, 34, 51, 27, and 35.

5. Round 209 to the nearest hundred.

6. $9 \times 2 = ?$

7. Which digit is in the thousands place in 7,421?

8. $916 - 363 = ?$

9. $0 \times 5 = ?$

10. Does a feather weigh more than or less than a pound?

11. Arrange these numbers from greatest to least. 917, 670, 232, 435

12. Write the standard number for five thousand, five hundred sixty-three.

13. Is your desk 3 inches tall or 3 feet tall?

14. What will be the time 25 minutes after 1:00?

15. How much money is shown?

1.	2.	3.
4.	5.	6.
7.	8.	9.
10.	11.	12.
13.	14.	15.

Summer Solutions© Mathematics   Level 2

Lesson #14

1. Draw two ovals that are <u>not</u> congruent.

2. The answer to a multiplication problem is the _____.

3. 145 + 639 = ?

4. Round 613 to the nearest hundred.

5. How much money do you have if you have 4 quarters?

6. 4 × 5 = ?

7. 710 − 356 = ?

8. How much time has passed between the two clocks?

9. 0 × 5 = ?

10. Which digit is in the ones place in 8,239?

11. What fraction is shaded?

12. 3,375 ◯ 4,257

13. 2 + 4 + 9 = ?

14. Count by tens. 260, 270, ____, ____, 300

15. How long is this piece of candy, in inches?

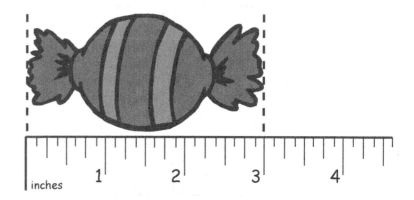

Summer Solutions© Mathematics — Level 2

1.	2.	3.
4.	5.	6.
7.	8.	9.
10.	11.	12.
13.	14.	15.

Lesson #15

1. 8 + 7 = ?

2. Are these cubes congruent?

3. Round 77 to the nearest ten.

4. 4 × 10 = ?

5. Write the time shown on the clock.

6. 830 − 216 = ?

7. List the even numbers between 31 and 36.

8. Write the standard number for six thousand, four hundred twenty-five.

9. What is the name of the shape?

10. How many inches are in a foot?

11. Which shaded fraction is greater?

    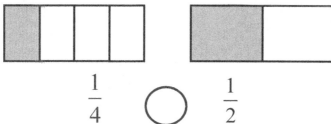

    $\frac{1}{4}$ ◯ $\frac{1}{2}$

12. 875 ◯ 1,279

13. What will be the time 15 minutes after 2:00?

14. Does a lamp weigh more than a pound or less than a pound?

15. Jason planted 19 seeds in one carton and 16 seeds in another carton. How many seeds did Jason plant in all?

Summer Solutions© Mathematics — Level 2

1.	2.	3.
4.	5.	6.
7.	8.	9.
10.	11.	12.
13.	14.	15.

Lesson #16

1. Is 3,528 an even number or an odd number?

2. How much money is shown?

3. 446 + 368 = ?

4. Put the numbers in order from least to greatest. 57, 10, 29, 19, 41

5. Is the line a line of symmetry?

6. 5 × 2 = ?

7. Which digit is in the tens place in 7,902?

8. Draw a square and shade $\frac{1}{4}$ of it.

9. How many minutes are in a half-hour?

10. Is this figure a cube?

11. 571 − 327 = ?

12. The _____ is the answer to a subtraction problem.

13. Tommy has 6 trees in his yard. There are 2 robins in each tree. How many robins are in Tommy's yard?

14. Round 695 to the nearest hundred.

15. March 1st will be on what day of the week?

    What is the date 3 weeks after February 3rd?

February

S	M	T	W	T	F	S
						1
2	3	4	5	6	7	8
9	10	11	12	13	14	15
16	17	18	19	20	21	22
23	24	25	26	27	28	

Summer Solutions© Mathematics — Level 2

1.	2.	3.
4.	5.	6.
7.	8.	9.
10.	11.	12.
13.	14.	15.

Lesson #17

1. Write 931 using words.

2. What number comes right before 700?

3. 78¢ + 35¢ = ?

4. Fill in the missing numbers. 425, ____, 445, ____

5. How much time has passed from one clock to the other?

6. 4 × 10 = ?

7. 769 − 361 = ?

8. What fraction is <u>not</u> shaded?

9. The sum is the answer to a(n) _____ problem.

10. In the set of numbers, 28, 37, 26, 37, and 61, what is the mode?

11. 558 + 262 = ?

12. How much money is shown here?

13. Which is longer, 9 inches or 1 foot?

14. 12 ÷ 2 = ?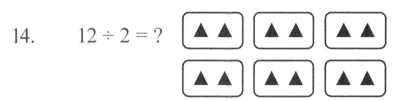

15. Does an envelope weigh more than a pound or less than a pound?

Summer Solutions© Mathematics — Level 2

1.	2.	3.
4.	5.	6.
7.	8.	9.
10.	11.	12.
13.	14.	15.

Lesson #18

1. Which month comes before March?

2. 8,422 ◯ 6,326

3. What is the name of the shape?

4. Write the next four odd numbers. 1, ____, ____, ____, ____

5. 824 + 323 = ?

6. 5 × 5 = ?

7. Round 688 to the nearest hundred.

8. 3 × 2 = ?

9. Draw rectangle and show a line of symmetry on it.

10. Does a CD weigh more than or less than a pound?

11. Which digit is in the ones place in 926?

12. 480 − 238 = ?

13. I have six thousands, four hundreds, two tens and six ones. What number am I?

14. Which even numbers come between 21 and 25?

15. How long does lunch last?

    Which activity lasts the longest?

Summer Schedule	
Time	Activity
12:30–1:00	Lunch
1:00– 2:00	Swimming Lessons
2:00–3:30	Soccer Practice
3:30–4:30	Chores

Summer Solutions© Mathematics Level 2

1.	2.	3.
4.	5.	6.
7.	8.	9.
10.	11.	12.
13.	14.	15.

Lesson #19

1. 53 + 39 = ?

2. How much money is 9 quarters?

3. Draw 2 congruent triangles.

4. Which is longer, 30 minutes or 1 hour?

5. 825 − 347 = ?

6. Put the numbers in order from greatest to least.

    2,158   7,513   1,521

7. 7 × 2 = ?

8. What fraction of the pie is shaded?

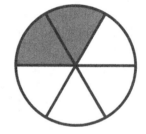

9. How many quarts are in a gallon?

10. What number comes between 265 and 450?

    145   567   358

11. Which digit is in the hundreds place in 8,312?

12. Write the name of the shape.

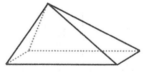

13. Is 217 closer to 200 or to 300?

14. What will be the time 10 minutes after 4:00?

15. Monica bought a notebook for $2.15 and a sticker book for $1.45. How much money did she spend?

Summer Solutions© Mathematics Level 2

1.	2.	3.
4.	5.	6.
7.	8.	9.
10.	11.	12.
13.	14.	15.

Lesson #20

1. Which digit is in the thousands place in 7,039?

2. Round 512 to the nearest hundred.

3. 664 + 125 = ?

4. What fraction of the pie is shaded?

5. 6 × 2 = ?

6. There are ___ inches in a foot.

7. Write the name of the figure.

8. 796 ◯ 988

9. Are these shapes congruent?

10. 65¢ − 32¢ = ?

11. How much time has passed between the two clocks?

12. Write the standard number for nine thousand, five hundred ninety-six.

13. Which shaded part is larger?

14. 890 − 347 = ?

    $\frac{1}{6}$ ◯ $\frac{1}{4}$

15. Charles has 55¢. His brother gives him 2 dimes and then he buys a piece of gum for 15¢. How much money does he have now?

Summer Solutions© Mathematics   Level 2

1.	2.	3.
4.	5.	6.
7.	8.	9.
10.	11.	12.
13.	14.	15.

Summer Solutions© Mathematics   Level 2

Lesson #21

1. $7.45 − $3.36 = ?

2. Draw a rectangle and shade $\frac{3}{4}$ of it.

3. Round 768 to the nearest hundred.

4. Put the numbers in order from least to greatest.

   913   1,341   655   2,247

5. 361 + 644 = ?

6. Does this figure show a line of symmetry?

7. If Juan buys a ball for 55¢ and pays for it with 3 quarters, how much change will he get back?

8. Write 294 using words.

9. Two shapes that are the same size and shape are _____.

10. Jason has 3 fish bowls in his room. There are 5 fish in each bowl. How many fish does Jason have?

11. In the number 951 which digit is in the tens place?

12. 11 − 6 = ?

13. 3 × 2 = ?

14. How much money is 5 dimes?

15. Which thermometer shows a good temperature for swimming?

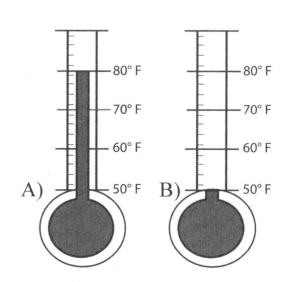

Summer Solutions© Mathematics — Level 2

1.	2.	3.
4.	5.	6.
7.	8.	9.
10.	11.	12.
13.	14.	15.

Lesson #22

1. 27 + 18 = ?

2. Count the coins that are shown here.

3. Round 532 to the nearest hundred.

4. The number that comes up most often in a set is the _____.

5. 765 − 372 = ?

6. Write the even numbers between 51 and 57.

7. I have 8 hundreds, 2 tens, and 5 ones. What number am I?

8. What is the name of this shape?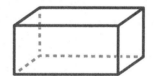

9. 546 + 168 = ?

10. 4 × 5 = ?

11. Which digit is in the hundreds place in 2,015?

12. How many days are in a year?

13. 8 ÷ 2 = ?

14. 9,642 ◯ 9,624

15. How long, in inches, is the paper clip?

Summer Solutions© Mathematics   Level 2

1.	2.	3.
4.	5.	6.
7.	8.	9.
10.	11.	12.
13.	14.	15.

Lesson #23

1. 2 × 5 = ?

2. Round 327 to the nearest hundred.

3. 92 − 57 = ?

4. Sam saves newspapers. He has five stacks, and each stack has 10 newspapers in it. How many newspapers has Sam saved?

5. Is 9,067 an even or an odd number?

6. 156 + 385 = ?

7. Are these shapes congruent?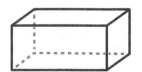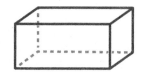

8. 0 × 10 = ?

9. What fraction of the pie is not shaded?

10. Count by fives.   55, ____, ____, 70

11. The answer to a multiplication problem is the _____.

12. Does a baby weigh more than a pound or less than a pound?

13. What time is shown on the clock?

14. Which is greater, 9 dimes or $1.00?

15. Taylor picked some pears and has put them in bags. She has 5 bags, with 5 pears in each bag. How many pears did Taylor pick?

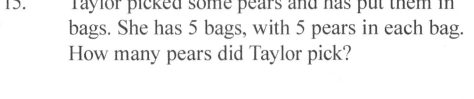

Summer Solutions© Mathematics   Level 2

1.	2.	3.
4.	5.	6.
7.	8.	9.
10.	11.	12.
13.	14.	15.

Lesson #24

1. 618 + 256 = ?

2. Does a piano weigh more than or less than a pound?

3. Draw two congruent rhombuses (diamonds).

4. Round 378 to the nearest hundred.

5. Write the name of the shape.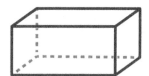

6. 802 − 571 = ?

7. The difference is the answer to a(n) _____ problem.

8. 10 ÷ 2 = ?

9. 8 + 7 = ?

10. How much time has passed from one clock to the next?

11. 0 × 5 = ?

12. 56 is the _____ of these numbers. 27, 29, 56, 43, 56

13. 7,412 ◯ 8,831

14. What fraction of the pie is shaded?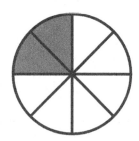

15. What is the length of the chalk in inches?

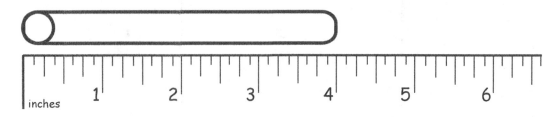

Summer Solutions© Mathematics　　　　　　　　　　　　　　　Level 2

1.	2.	3.
4.	5.	6.
7.	8.	9.
10.	11.	12.
13.	14.	15.

Lesson #25

1. Write 462 using words.

2. What is the name of this shape?

3. Which digit is in the tens place in 9,745?

4. Write a fact family for 3, 8, and 11.

5. Count the coins.

6. 834 − 672 = ?

7. Would you measure the weight of a dog in feet, pounds, or in quarts?

8. How many minutes are in a half-hour?

9. 638 + 259 = ?

10. Which shape is divided into two equal parts? Draw it.

11. Write the odd numbers between 50 and 56.

12. What time is shown on the clock?

13. 5 × 5 = ?

14. Does it take 20 minutes or 20 hours to cut the grass?

15. How many total animals were in the barnyard?

    How many more chickens than cows were there?

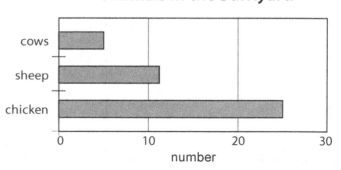

Summer Solutions© Mathematics — Level 2

1.	2.	3.
4.	5.	6.
7.	8.	9.
10.	11.	12.
13.	14.	15.

Lesson #26

1. I have four thousands, three hundreds, five tens, and seven ones. What number am I?

2. Write the name of the shape.

3. 45 + 31 + 12 = ?

4. Are these shapes congruent?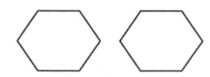

5. The answer to a(n) addition problem is the _____.

6. How much time has passed between the two clocks?

7. 620 − 338 = ?

8. Is the number 4,615 even or odd?

9. Eight quarters is the same amount of money as two _____.

10. 9 × 2 = ?

11. Order these numbers from greatest to least.

    555   1,141   783   248

12. Round 67 to the nearest ten.

13. 5,452 ◯ 5,524

14. Write 916 using words.

15. How many days are in July?

    August 1st will be on what day of the week?

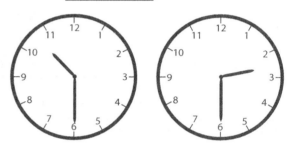

1.	2.	3.
4.	5.	6.
7.	8.	9.
10.	11.	12.
13.	14.	15.

Lesson #27

1. There are _____ quarts in a gallon.

2. 853 + 197 = ?

3. What is the name of this shape?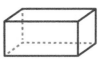

4. Which digit is in the hundreds place in 2,463?

5. What number comes just before 900?

6. What fraction of the rectangle is shaded?

7. Draw 2 squares that are not congruent.

8. Write the time shown on the clock.

9. 814 − 436 = ?

10. Marcus bought a football for $4.25. He gave the clerk $5.00. How much change did he receive?

11. Write a fact family for 6, 7, and 13.

12. How much money is shown here?

13. 6 × 5 = ?

14. Order the numbers from least to greatest.

    3,385   2,279   3,028   2,436

15. What temperature is shown on the thermometer?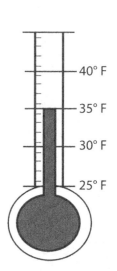

Summer Solutions© Mathematics — Level 2

1.	2.	3.
4.	5.	6.
7.	8.	9.
10.	11.	12.
13.	14.	15.

Lesson #28

1. Round 758 to the nearest hundred.

2. 943 − 385 = ?

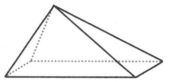

3. Write the name of this shape.

4. Draw 2 congruent triangles.

5. A number has 5 thousands, 4 hundreds, 5 tens, and 3 ones. What is the number?

6. 2,318 ◯ 2,138

7. Which would you use to measure the temperature?

       A) scale   B) ruler   C) thermometer

8. How many hours are in a day?

9. 645 + 291 = ?

10. How much money is shown?

11. 6 × 2 = ?

12. Is the number 6,524 even or odd?

13. Does this show a line of symmetry?

14. Which is longer, 3 weeks or 1 month?

15. How long is the feather, in inches?

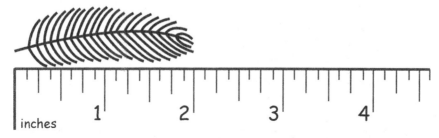

Summer Solutions© Mathematics   Level 2

1.	2.	3.
4.	5.	6.
7.	8.	9.
10.	11.	12.
13.	14.	15.

Lesson #29

1. Which digit is in the ones place in 4,579?

2. 475 + 332 = ?

3. Is this figure a cone?

4. Count by tens. 650, 660, _____, _____, _____

5. Round 214 to the nearest hundred.

6. What time is it on this clock?

7. $2.65 − $1.53 = ?

8. 7,314 ◯ 7,143

9. Does it take 7 seconds or 7 minutes to get dressed?

10. What fraction is shaded?

11. 4 × 5 = ?

12. The number that comes up most often in a set is the _____.

13. Twenty nickels is the same amount of money as one _____.

14. Are these shapes congruent?

    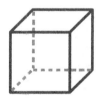   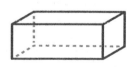

15. There are 58 students in the school band. Twenty-six students play a wind instrument. How many students <u>do not</u> play a wind instrument?

Summer Solutions© Mathematics — Level 2

1.	2.	3.
4.	5.	6.
7.	8.	9.
10.	11.	12.
13.	14.	15.

Lesson #30

1. 46 − 18 = ?

2. Which digit is in the tens place in the number 571?

3. Jessie planted 3 rows of tomato plants with 5 plants in each row. How many tomato plants did Jessie plant?

4. Round 26 to the nearest ten.

5. 814 + 357 = ?

6. List the odd numbers between 71 and 77.

7. Write the missing numbers in the sequence.   25, 35, _____, 55, _____

8. Is this figure a cone?

9. 5 × 10 = ?

10. If you have 8 dimes and 4 nickels, how much money do you have?

11. Two figures with the same size and shape are _____.

12. How many months are in two years?

13. Which shaded part is greater?

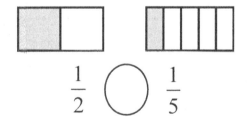

$\frac{1}{2}$ ◯ $\frac{1}{5}$

14. 1 × 9 = ?

15. What will be the time 30 minutes after 3:00?

1.	2.	3.
4.	5.	6.
7.	8.	9.
10.	11.	12.
13.	14.	15.

Level 2

Mathematics
3rd Edition

Help Pages

Help Pages

Vocabulary

Arithmetic Operations
addition — the values of two or more numbers joined together. The sign "+" means add. The answer to an addition problem is called the **sum**. Example: Put the values of 5 and 2 together; the sum is 7; 5 + 2 = 7.
subtraction — a value taken away from another. The sign "−" means subtract. The answer to a subtraction problem is called the **difference**. Example: Take 1 away from 5, the difference is 4; 5 − 1 = 4.
multiplication — a value added to itself repeatedly. The sign "×" means multiply. The answer to a multiplication problem is called the **product**. Example: When 5 is added to itself 3 times, the product is 15; 5 + 5 + 5 is the same as 3 × 5 = 15.
division — a value shared equally. The sign "÷" means divide. The answer to a division problem is called the **quotient**. Example: When 8 is shared equally between 2, the quotient is 4; 8 ÷ 2 = 4.

Geometry
congruent — figures with the same shape and the same size.
fraction — a part of a whole. Example: This box has 4 parts. 1 part is shaded; $\frac{1}{4}$.
line of symmetry — a line along which a figure can be folded so that the two halves match exactly.

Geometry — Shapes and Solids			
cone	△	pyramid	
cube		rectangular prism	
cylinder		rhombus (diamond)	
ellipse (oval)		sphere	

Summer Solutions© Mathematics   Level 2

Help Pages

Vocabulary

Geometry — Polygons				
Number of Sides	Name	Number of Sides	Name
3	△ triangle	4	▭ quadrilateral

Measurement — Relationships	
Time	**Distance**
30 minutes = 1 half-hour	12 inches = 1 foot
60 minutes = 1 hour	**Volume**
365 days = 1 year	4 quarts = 1 gallon

Statistics

mode – the number that occurs most often in a group of numbers. The mode is found by counting how many times each number occurs in a list. The number that occurs more than any other is the mode. Some groups of numbers have more than one mode.

Example: The mode of 77, ⓷93, 85, ⓷93, 77, 81, ⓷93 and 71 is **93**.
(93 is the mode because it occurs more than the others.)

Place Value

Whole Numbers

1, 4 0 5

Thousands Hundreds Tens Ones

The number above is read: one thousand, four hundred five.

65

Help Pages

Solved Examples

Whole Numbers (continued)

Rounding numbers means estimating them. Focus on a particular place value, and decide if that digit is closer to the next higher number (round up) or to the next lower number (keep the same). It might be helpful to look at the place-value chart in the Help Pages.

Example: Round 347 to the tens place.

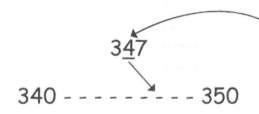

347 is closer to 350, so it is rounded to 350.

1. Identify the place to round to.
2. What are the nearest "tens" on either side of the number? (340 and 350)
3. Which of these is 347 closer to?
4. This is the number to round to.

350

Here is another example of rounding whole numbers.

Example: Round 83 to the nearest ten.

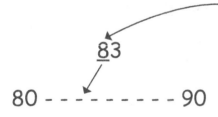

83 is closer to 80, so it is rounded to 80.

1. What is the rounding place?
2. What are the nearest "tens" on either side of the number? (80 and 90)
3. Which of these is 83 closer to?
4. This is the number to round to.

80

Help Pages

Solved Examples

Whole Numbers (continued)

There are **even numbers** and **odd numbers**. A number is <u>even</u> if it ends in 0, 2, 4, 6, or 8. A number is <u>odd</u> if it ends in 1, 3, 5, 7, or 9.

Examples: 46 is an even number because it ends in 6.

11 is an odd number because it ends in 1.

A **fact family** is a set of related facts using addition, subtraction, and the same three numbers.

Example: Write a fact family using 3, 4, and 7.

$$3 + 4 = 7 \qquad 7 - 3 = 4$$
$$4 + 3 = 7 \qquad 7 - 4 = 3$$

Numbers can be compared by saying one is **greater than** another or one is **less than** another.

The symbol ">" means *greater than*. The symbol "<" means *less than*. (Hint: The open part of the sign is near the bigger number.)

Examples: 10 < 18   10 is less than 18.

27 > 13   27 is greater than 13.

Summer Solutions® Mathematics   Level 2

Help Pages

Solved Examples

Whole Numbers (continued)

When adding or subtracting whole numbers, the numbers must first be lined-up from the right. Starting with the ones place, add (or subtract) the numbers. When adding, if the answer has 2 digits, write the ones digit and regroup the tens digit. For subtraction, it may also be necessary to regroup first. Then, add (or subtract) the numbers in the tens place. Continue with the hundreds, etc.

Look at these examples of **addition**.

Examples: Find the sum of 314 and 12.   Add 648 and 236.

```
  314                                              ¹
+  12                                            648
-----                                          + 236
  326                                          -----
                                                 884
```

1. Line up the numbers on the right.
2. Beginning with the ones place, add. Regroup if necessary.
3. Repeat with the tens place.
4. Continue this process with the hundreds place, etc.

Use the following examples of **subtraction** to help.

Example: Subtract 37 from 93.

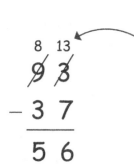

1. Begin with the ones place. Check to see if regrouping is needed. Since 7 is larger than 3, regroup to 8 tens and 13 ones.
2. Now look at the tens place. Check to see if regrouping is needed. Since 3 is less than 8, regrouping is not needed.
3. Subtract each place value beginning with the ones.

Help Pages

Solved Examples

Whole Numbers (continued)

Example: Find the difference of 425 and 233.

```
  3 12
  4 2 5
- 2 3 3
―――――――
  1 9 2
```

1. Begin with the ones place. Check to see if regrouping is needed. Since 3 is less than 5, regrouping is not needed.
2. Now look at the tens place. Check to see if regrouping is needed. Since 3 is larger than 2, regroup to 3 hundreds and 12 tens.
3. Now look at the hundreds place. Check to see if regrouping is needed. Since 2 is less than 3, subtract.
4. Subtract each place value beginning with the ones.

When **subtracting from zero**, regrouping is always needed. Use the examples below to help.

Example: Subtract 38 from 60.

```
  5 10
  6 0
- 3 8
―――――
  2 2
```

1. Begin with the ones place. Since 8 is less than 0, regrouping is needed.
2. Regroup to 5 tens and 10 ones.
3. Then, subtract each place value beginning with the ones.

Example: Find the difference between 500 and 261.

```
      9
  4 10 10
  5 0 0
- 2 6 1
―――――――
  2 3 9
```

Help Pages

Solved Examples

Whole Numbers (continued)

Multiplication is a quick way to add groups of numbers. The sign (×) for multiplication is read "times." The answer to a multiplication problem is called the **product**. Use the examples below to help understand multiplication.

Example: 2 × 5 is read "two times five."

    It means *2 groups of 5* or 5 + 5.

2 × 5 = 5 + 5 = 10

The product of 2 × 5 is 10.

Example: 5 × 4 is read "five times four."

    It means *5 groups of 4* or 4 + 4 + 4 + 4 + 4.

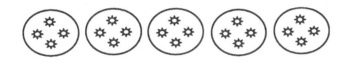

5 × 4 = 4 + 4 + 4 + 4 + 4 = 20

The product of 5 × 4 is 20.

It is very important to memorize **multiplication facts**. This table will help, but only until it's memorized!

To use this table, choose a number in the top gray box and multiply it by a number in the left gray box. Follow both with a finger (down and across) until they meet. The number in that box is the product.

An example is shown.

    2 × 5 = 10

×	0	1	2	5	10
0	0	0	0	0	0
1	0	1	2	5	10
2	0	2	4	10	20
5	0	5	10	25	50
10	0	10	20	50	100

Help Pages
Solved Examples

Whole Numbers (continued)

Division is the opposite of multiplication. The sign (÷) for division is read "divided by." The answer to a division problem is called the **quotient**.

Remember that multiplication is a way of adding groups to get their total. Think of division as the opposite of this. In division, the total and the number in each group are already known. The total number of groups is unknown. Follow the examples below.

Example: What is 9 ÷ 3?   (9 items divided into groups of 3.)

The total number is 9.

Each group contains 3.

How many groups are there?   There are 3 groups.

9 ÷ 3 = **3**

Example: Divide 10 by 2.   (10 items divided into groups of 2.)

The total number is 10.

Each group contains 2.

How many groups are there?   There are 5 groups.

10 ÷ 2 = 5

Fractions

A **fraction** is used to represent part of a whole. The top number in a fraction is the part. The bottom number in a fraction is the whole.

The whole rectangle has 6 sections.

Only 1 section is shaded.

This can be shown as the fraction $\frac{1}{6}$.

$\frac{1}{6}$ $\frac{\text{shaded part}}{\text{parts in the whole}}$

Help Pages

Solved Examples

Time

The measure of how long something takes to happen is called **elapsed time**.

Example:

The movie began at 7:00  and ended at 9:00 .

How long did the movie last? (How much time passed between 7:00 and 9:00?) There are **2 hours** between 7:00 and 9:00, so the movie lasted for 2 hours.

Example:

How many hours pass from the beginning of Spelling class until the end of Math class?

Class Schedule

8:30 – 9:00	Spelling
9:00 – 10:00	Reading
10:00 – 11:30	Math
11:30 – 12:00	English

Spelling starts at 8:30. Math ends at 11:30. (How much time passes between 8:30 and 11:30?)

There are **3 hours** between 8:30 and 11:30. Three hours pass from the beginning of Spelling class until the end of Math class.

Who Knows?

Sides in a triangle?..(3)

Sides in a square?..(4)

Days in a week?...(7)

Months in a year?..(12)

Days in a year?..(365)

Inches in a foot?..(12)

Quarts in a gallon?...(4)

The number that is seen most
often in a set of numbers?..(mode)

Figures with the same size
and shape?..(congruent)

Answer to an addition problem?............................(sum)

Answer to a subtraction problem?............(difference)

Answer to a multiplication problem?..............(product)

Summer Solutions© Mathematics — Level 2

Level 2

Mathematics
3rd Edition

Answers to Lessons

	Lesson #1		Lesson #2		Lesson #3
1	741	1	9:35	1	5
2	○ ○	2	112	2	yes
3	6:20	3	20 wings	3	940
4	301	4	27	4	−
5	×	5	even number	5	15 plants
6	4 quarts	6	cube	6	58, 60, 64
7	4	7	three hundred forty-one	7	40
8	less than	8	12	8	☐ ☐
9	<	9	0	9	more than
10	$\frac{1}{4}$	10	$1.75	10	cone, cylinder, pyramid
11	27	11	373	11	1 hour, 30 minutes
12	$2.36	12	multiplication	12	15
13	2	13	7	13	65
14	$3.15	14	0	14	32, 34
15	5 inches	15	50°F	15	5 inches

	Lesson #4		Lesson #5		Lesson #6
1	3 dimes	1	1 month	1	32¢
2	>	2	71	2	4 + 5 = 9 9 − 4 = 5 5 + 4 = 9 9 − 5 = 4
3	1, 3, 5, 7, 9	3	rhombus (diamond)	3	♡ ♡
4	974	4	>	4	odd
5	40¢	5	congruent	5	931
6	sum	6	4	6	25
7	7	7	10	7	no
8	more than	8	6:20	8	8
9	cone	9	314	9	$\frac{1}{3}$
10	8:55	10	2	10	>
11	495	11	$\frac{2}{5}$	11	6 feet long
12	40	12	4 quarts	12	280, 300
13	c) 205	13	73¢	13	70
14	0	14	15	14	12 inches
15	35 fish; 10 more	15	15 rooms	15	Thursday June 20th

	Lesson #7		Lesson #8		Lesson #9
1	42, 44	1	16	1	31
2	316	2	$\frac{3}{6}$	2	1:25
3	$4.32	3	20 people	3	50
4	60 minutes	4	24 months	4	<
5	12	5	pyramid	5	1,347
6	12:45	6	$2\frac{1}{2}$ hours	6	12 inches
7	151	7	995	7	7
8	730 days	8	71, 73, 75	8	6
9	cylinder	9	3	9	$1.00
10	10	10	multiplication	10	500
11	difference	11	211	11	odd
12	November	12	$\frac{1}{4}$	12	4
13	999	13	$2.00	13	5 inches
14	seven hundred ninety-seven	14	40	14	
15	393 more people	15	Reading; 1 hour; 10:00	15	70°F

	Lesson #10		Lesson #11		Lesson #12
1	1 year	1	mode	1	40
2	807	2	35	2	$\frac{3}{8}$
3	○ ○	3	400	3	$1.50
4	$2.56	4	$1.22	4	11
5	0	5	cone	5	$1\frac{1}{2}$ hours
6	300	6	66, 70, 72	6	751
7	364	7	82, 84, 86	7	30
8	3:55	8	$\frac{1}{6}$	8	addition
9	>	9	678	9	pyramid
10	no	10	8	10	383
11	0	11	no	11	>
12	774	12	50, 60, 65	12	300
13	$\frac{2}{6}$	13	7:50	13	5
14	325, 465, 787, 1,254	14	$4+9=13$ $13-4=9$ $9+4=13$ $13-9=4$	14	1, 3, 5, 7
15	10 children; 3 more	15	33 cookies	15	4 inches

	Lesson #13		Lesson #14		Lesson #15
1	882	1	⬭ ○	1	15
2	▨▨▨☐	2	product	2	no
3	<	3	784	3	80
4	27	4	600	4	40
5	200	5	$1.00	5	6:40
6	18	6	20	6	614
7	7	7	354	7	32, 34
8	553	8	$2\frac{1}{2}$ hours	8	6,425
9	0	9	0	9	sphere
10	less than	10	9	10	12 inches
11	917, 670, 435, 232	11	$\frac{3}{6}$	11	$\frac{1}{2}$ <
12	5,563	12	<	12	<
13	3 feet	13	15	13	2:15
14	1:25	14	280, 290	14	more than
15	$3.17	15	3 inches	15	35 seeds

	Lesson #16		Lesson #17		Lesson #18
1	even	1	nine hundred thirty-one	1	February
2	$1.75	2	699	2	>
3	814	3	$1.13	3	cylinder
4	10, 19, 29, 41, 57	4	435, 455	4	3, 5, 7, 9
5	yes	5	3 hours	5	1,147
6	10	6	40	6	25
7	0	7	408	7	700
8		8	$\frac{1}{3}$	8	6
9	30 minutes	9	addition	9	
10	no	10	37	10	less than
11	244	11	820	11	6
12	difference	12	51¢	12	242
13	12 robins	13	1 foot	13	6,426
14	700	14	6	14	22, 24
15	Saturday, February 24th	15	less than	15	30 minutes; Soccer Practice

	Lesson #19		Lesson #20		Lesson #21
1	92	1	7	1	$4.09
2	$2.25	2	500	2	▨▨▨☐
3	△ △	3	789	3	800
4	1 hour	4	$\frac{4}{5}$	4	655; 913; 1,341; 2,247
5	478	5	12	5	1,005
6	7,513; 2,158; 1,521	6	12	6	no
7	14	7	sphere	7	20¢
8	$\frac{2}{6}$	8	<	8	two hundred ninety-four
9	4 quarts	9	no	9	congruent
10	358	10	33¢	10	15 fish
11	3	11	$1\frac{1}{2}$ hours	11	5
12	pyramid	12	9,596	12	5
13	200	13	$\frac{1}{4}$ <	13	6
14	4:10	14	543	14	50¢
15	$3.60	15	60¢	15	A) 80°F

	Lesson #22		Lesson #23		Lesson #24
1	45	1	10	1	874
2	51¢	2	300	2	more than
3	500	3	35	3	◇ ◇
4	mode	4	50 newspapers	4	400
5	393	5	odd number	5	rectangular prism
6	52, 54, 56	6	541	6	231
7	825	7	yes	7	subtraction
8	rectangular prism	8	0	8	5
9	714	9	$\frac{3}{4}$	9	15
10	20	10	60, 65	10	4 hours
11	0	11	product	11	0
12	365 days	12	more than	12	mode
13	4	13	8:45	13	<
14	>	14	$1.00	14	$\frac{2}{8}$
15	1 inch	15	25 pears	15	4 inches

	Lesson #25		Lesson #26		Lesson #27
1	four hundred sixty-two	1	4,357	1	4
2	sphere	2	cube	2	1,050
3	4	3	88	3	rectangular prism
4	$3+8=11$ $11-3=8$ $8+3=11$ $11-8=3$	4	yes	4	4
5	37¢	5	sum	5	899
6	162	6	4 hours	6	$\frac{1}{4}$
7	pounds	7	282	7	☐ ☐
8	30 minutes	8	odd	8	6:15
9	897	9	dollars	9	378
10	⟵⊖⟶	10	18	10	75¢
11	51, 53, 55	11	1,141; 783; 555; 248	11	$6+7=13$ $13-6=7$ $7+6=13$ $13-7=6$
12	10:20	12	70	12	$4.35
13	25	13	<	13	30
14	20 minutes	14	nine hundred sixteen	14	2,279; 2,436; 3,028; 3,385
15	about 42 animals; 20 more	15	31 days; Saturday	15	35°F

	Lesson #28		Lesson #29		Lesson #30
1	800	1	9	1	28
2	558	2	807	2	7
3	pyramid	3	no	3	15 tomato plants
4	△ △	4	670, 680, 690	4	30
5	5,453	5	200	5	1,171
6	>	6	11:40	6	73, 75
7	C) thermometer	7	$1.12	7	45, 65
8	24 hours	8	>	8	no
9	936	9	7 minutes	9	50
10	$2.85	10	$\frac{1}{2}$	10	$1.00
11	12	11	20	11	congruent
12	even	12	mode	12	24 months
13	yes	13	dollar	13	$\frac{1}{2}$ >
14	1 month	14	no	14	9
15	2 inches	15	32 students	15	3:30